Body Systems

JNF
611
C66b
1984

Written by **Lorraine Conway**

Illustrations by **Linda Akins**

Cover by Kathryn Hyndman

GOOD APPLE, INC.
BOX 299
CARTHAGE, IL 62321

Copyright © Good Apple, Inc., 1984

ISBN No. 0-86653-153-X

Printing No. # 987654

The purchase of this book entitles the buyer to reproduce student activity pages for classroom use only. Any other use requires written permission from Good Apple, Inc.

**GOOD APPLE, INC.
BOX 299
CARTHAGE, IL 62321**

Table of Contents

Activity   Page

Systems of the Body 1
The Skeletal System 2
The Skeleton 3
What Is Bone? 5
Parts of a Bone 6
Drawing the Skeleton 7
Determine the Fracture 8
Joints 10
The Muscular System 12
The Muscular System Front View 13
The Muscular System Back View 14
Muscles Move Bones 15
The Circulatory System 17
Arteries, Veins, and Capillaries 18
Why Is Blood Red? 19
Circulation of Blood Through the Heart 20
Make a Pulsemeter 21
The Respiratory System 22
A Model of the Respiratory System 24
How Much Air Can Your Lungs Hold? 25
How Much Oxygen Is Left After Breathing? 26
The Digestive System 27
Breaking Down Food for Digestion 29
Teeth 30

The Parts of a Tooth	31
The Excretory System	32
A Test for Diabetes	33
The Nervous System	34
The Brain	36
Our Sense of Sight	37
The Eye	38
Eye Tricks	39
Test Your Stereoscopic Vision	40
Our Sense of Hearing	41
The Ear	42
Using One Ear to Detect Sounds	43
Our Sense of Touch	44
How Well Do You Feel?	45
Hot or Cold?	46
Our Sense of Taste	47
Rate the Taste	48
Our Sense of Smell	49
Is It Taste or Smell?	50
Warning Smells	51
Anatomy of the Mouth, Nose and Throat	52
The Female Reproductive System	53
The Male Reproductive System	54
The Endocrine System	55
Endocrine Glands	56
Answer Key	57

Introduction

Body Systems offers an explanation and drawing of each of the major systems of the human body and of the five senses. Each system is coordinated with activities, demonstrations or experiments which are designed to involve students in acquiring knowledge concerning the structure and function of their bodies in the best possible way, through experiences, which ultimately are the best teachers.

The drawings of the body systems can be used in several ways: as preview lessons to determine the students prior knowledge of the subject, as learning guides, or, with the answers removed, as testing materials. The system explanations provide easy-to-understand introductions as the activities greatly enhance the students' knowledge and curiosity in this popular area of study.

Personal Teaching Notes

Systems of the Body

The human body is made of several systems working together to keep it alive and healthy. The systems in turn are made up of organs. For example, the stomach is an organ of the digestive system. The major systems of the human body are listed in the first column. For each system fill in its major organs and functions.

System	Major Organs	Major Functions
Skeletal		
Muscular		
Circulatory		
Respiratory		
Digestive		
Excretory		
Nervous		
Reproductive		
Endocrine		

The Skeletal System

The skeletal system is made up of the 206 bones of the human body. The functions of the skeletal system are:

1. To give support to the body.
2. To protect the vital organs such as the heart, lungs and brain.
3. To serve as a place of attachment for the skeletal muscles which move the body.
4. To supply the body with certain blood cells.
5. To store minerals that the body can use in times of need.

The bones of the body vary greatly in size and shape. The three bones of the middle ear - the hammer, anvil and stirrup are so small all three could easily fit on your thumbnail. The longest bone of the body is the femur which reaches from the hip to the knee.

Most bones of the body are formed from cartilage. The process of bone cells replacing cartilage is called ossification. Ossification occurs mostly in youth but can occur in later years as in the knitting of broken bones. The specialized bone-forming cells are called osteoblasts and osteoclasts. These words come from the Greek word for bone, *osteo*.

Ossification cannot occur unless calcium compounds are present. These compounds are brought to the bone-forming cells by the blood which travels through passages in the bones called Haversian canals. The bone cells, blood and nerves are the living part of bone. The minerals which make the bones strong and tough are the nonliving part of bone.

The Skeleton

Common Names	Scientific Names
1. Top of the skull	1. Cranium
2. Lower Jaw	2. Mandible
3. Shoulder blade	3. Scapula
4. Collarbone	4. Clavicle
5. Breastbone	5. Sternum
6. Bone of the upper arm	6. Humerus
7. Rib	7. Rib
8. Floating rib	8. Floating rib
9. Bone of the lower arm	9. Radius
10. Bone of the lower arm	10. Ulna
11. Hip bone	11. Pelvis
12. Wrist bone	12. Carpal
13. Hand bone	13. Metacarpal
14. Finger bone	14. Phalange
15. Bone of the upper leg	15. Femur
16. Kneecap	16. Patella
17. Bone of the lower leg	17. Tibia
18. Bone of the lower leg	18. Fibula
19. Ankle bone	19. Tarsal
20. Foot bone	20. Metatarsal
21. Toe bone	21. Phalange
22. Backbone	22. Vertebra

Use the scientific names to label the skeleton on the next page. Learn the scientific names of the bones.

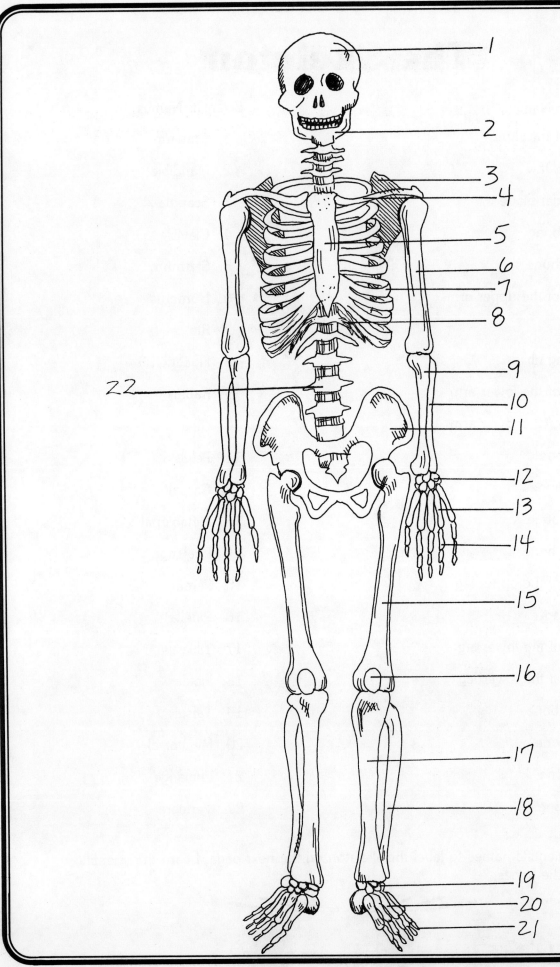

What Is Bone?

Bones are made of salts such as calcium phosphate and protein. The salts in bone give it strength; the protein helps to make the bone slightly elastic. The following experiments will help you understand more about bones.

MATERIALS: Two beakers, vinegar or 15% diluted hydrochloric acid, tongs, water, Bunsen burner, three chicken bones.

PROCEDURE: Place one chicken bone in a beaker of water, another in a beaker with either the dilute hydrochloric acid or vinegar. Allow to stand for several days. Using tongs, hold the third bone over a hot flame.

1. Explain what happened to the bone in the water.
2. What happened to the bone in the acid?
3. Did the acid affect the protein or the salts in the bone?
4. What happened to the chicken bone when it was held over the flame?
5. Did the salts or the protein burn?
6. Bones are made of living cells. Explain why a balanced diet is necessary for healthy bones.

Parts of a Bone

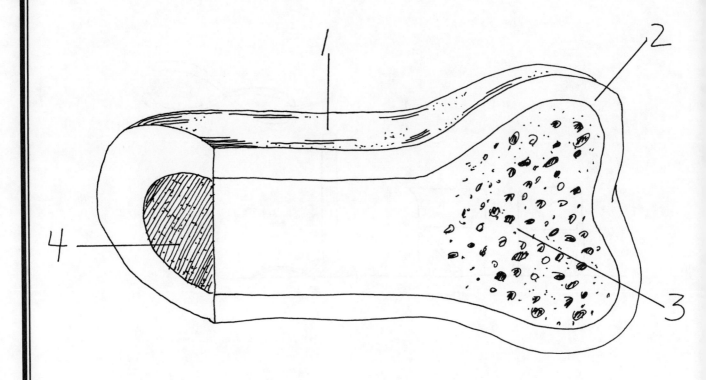

MATERIALS: Bones from the meat department of a grocery store or a butcher shop. Ask the butcher to cut the bone crosswise and lengthwise, or obtain two bones, one cut crosswise and the other cut lengthwise.

PROCEDURE: Thoroughly wash and rinse the bones before using, then find the following parts:

1. Covering or periosteum. This soft thin substance covers and protects the hard bone.
2. Hard bone. This is the tough and compact part of the bone; it can regrow when broken.
3. Spongy bone. This part of the bone often contains red marrow which makes red blood cells.
4. Marrow. This is the soft inner center of bones; it contains blood vessels and fat cells.

Drawing the Skeleton

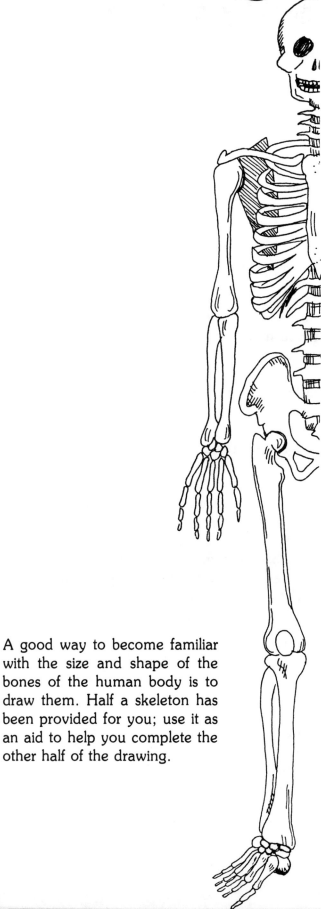

A good way to become familiar with the size and shape of the bones of the human body is to draw them. Half a skeleton has been provided for you; use it as an aid to help you complete the other half of the drawing.

Determine the Fracture

Dr. York, an orthopedic surgeon, sets many broken bones. Can you match the following descriptions of some of his patients' injuries to their x-rays on the following page?

1. When Karen fell from her horse, the result was a compound fracture of the tibia. In a compound fracture, part or parts of a bone break through the skin. Do you know which x-ray is Karen's?

2. Fred is a painter. He recently slipped from his ladder and broke his femur. His fracture is called an impacted fracture because one end of the broken bone was driven into the other broken end. Which x-ray is Fred's?

3. Ricky is only six years old. He fell from his skateboard and bent or curved the radius in his arm. Can you pick out Ricky's x-ray?

4. Mary slipped on a wet floor and suffered a simple fracture of the tibia. The bone did not move out of place. Find Mary's x-ray.

5. Fuller works in the post office. He tripped over a package and cracked his femur. It is called a partial fracture. Which is Fuller's x-ray?

6. Todd's motorcycle was hit by an automobile. His fracture was very severe because the bone was broken into many fragments. It is called a comminuted fracture of the femur. Which x-ray depicts a fragmented bone?

Determine the Fracture

The drawings below represent x-rays of Dr. York's patients. Can you match the x-rays to their descriptions on the preceding page? Write each patient's name under his or her x-ray.

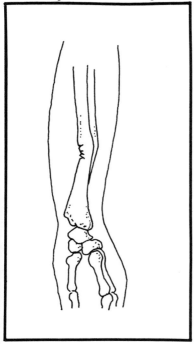

1. _____x-ray

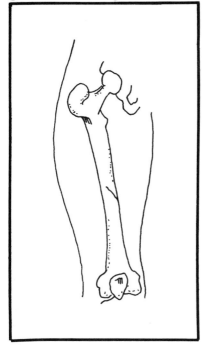

2. _____x-ray

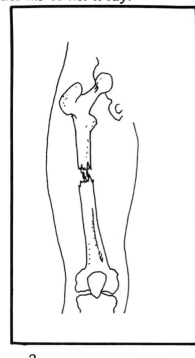

3. _____x-ray

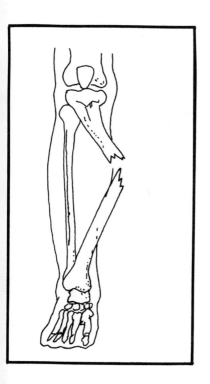

4. _____x-ray

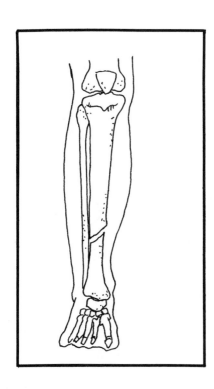

5. _____x-ray

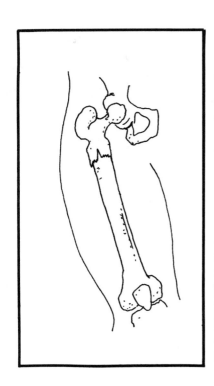

6. _____x-ray

Joints

Places where bones come together are called joints. Joints are held in place by tough bands of tissue called ligaments. The knees and elbows are examples of one type of joint called a hinge joint. These joints allow movement in only one direction similar to a door or hinge. In another kind of joint, one bone is ball-shaped and fits into the socket of another. Ball and socket joints are found in the hips and shoulders. These joints allow movements such as throwing a baseball and rotating the hip. The vertebrae of the back are slightly moveable joints which allow bending and twisting. In the skull are many immoveable joints. The eight bones of the wrist and seven bones of the ankle provide gliding joints where one bone can slide over another. The first two bones of the neck, the atlas and axis, form a pivot joint which allows movement in a circular motion as in the twisting of the head.

A special fluid called synovial (suh-nove-ee-ul) fluid lubricates the joints. In old age this fluid tends to dry up, thereby causing the squeaking bones of the elderly.

A good way to study joints is by obtaining x-rays which are no longer needed from doctors, chiropractors, clinics and veterinarians; another way is to identify the drawings on the following page by telling whether a hinge, ball and socket, slightly moveable, immoveable, gliding or pivot joint is represented.

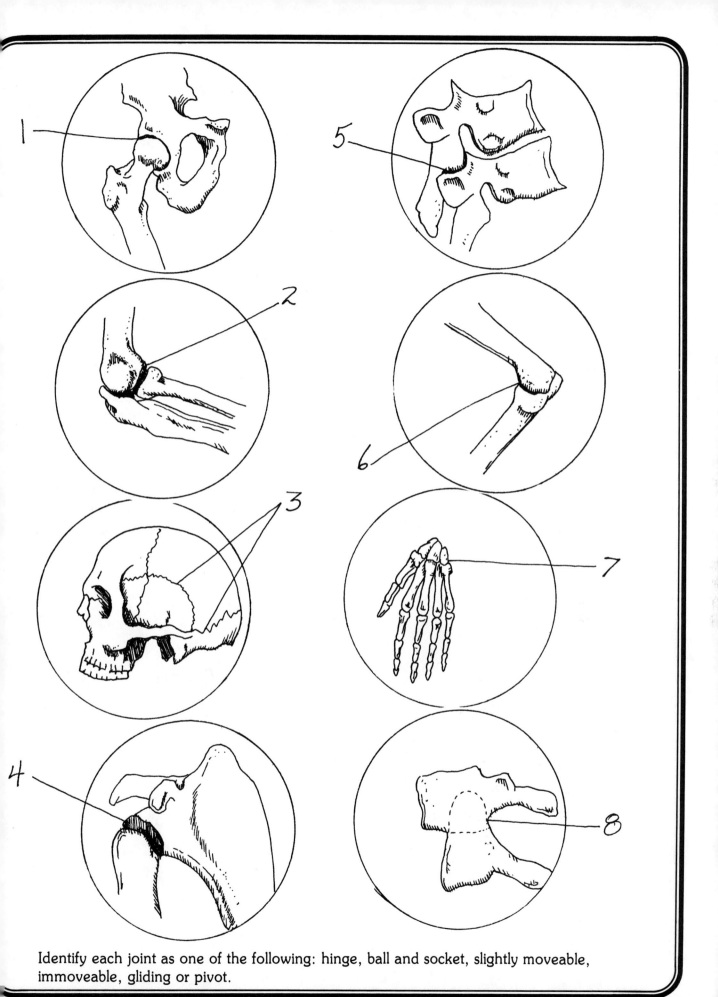

Identify each joint as one of the following: hinge, ball and socket, slightly moveable, immoveable, gliding or pivot.

The Muscular System

Almost half of the human body's weight is made up of muscle tissue of which there are three types:
1. Cardiac muscle is found only in the heart. It is involuntary; this means that it cannot be controlled at will.
2. Smooth muscle is found in the walls of the stomach and intestines and is also involuntary.
3. Skeletal muscles are attached to and move the bones of the skeleton. Most skeletal muscles are voluntary and can be moved at will.

Skeletal muscles are capable of moving bones because of their location and attachment. Most skeletal muscles are attached to bones by tendons. The end of the muscle that is attached to the bone that does not move is called the origin of the muscle. The part of the muscle that is attached to the moving bone is called the insertion of the muscle. The main part of the muscle is called the belly. Most skeletal muscles work in pairs as when the biceps muscle of the upper arm contracts to bend the arm, the opposite muscle, the triceps, is extended. When the triceps contracts to straighten the arm, the biceps is extended.

There are over 600 muscles in the human body. They have been given Latin names based on their shape, direction, location, number of origins (heads), function, or the bones to which they are attached. Can you match the names of the following muscles to their Latin meanings?

____1.   Biceps   ____a.   Chewing

____2.   Rectus femoris   ____b.   Two heads

____3.   Subscapularis   ____c.   Triangular

____4.   Masseter   ____d.   Under the scapular

____5.   Deltoid   ____e.   Straight of the femur

____6.   Tibialis anterior   ____f.   In front of the tibia

The Muscular System
Front View

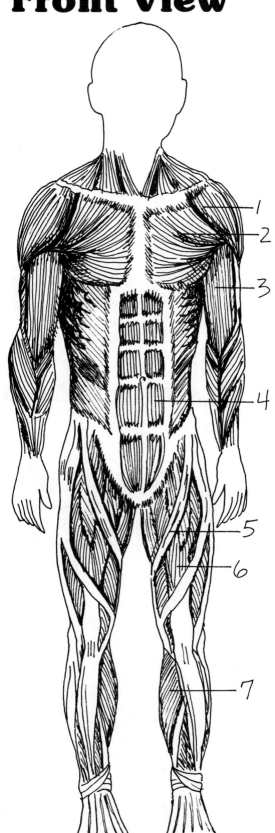

1. Deltoid
2. Pectoralis major
3. Biceps
4. Rectus abdominis
5. Sartorius
6. Quadriceps femoris
7. Gastrocnemius

The Muscular System
Back View

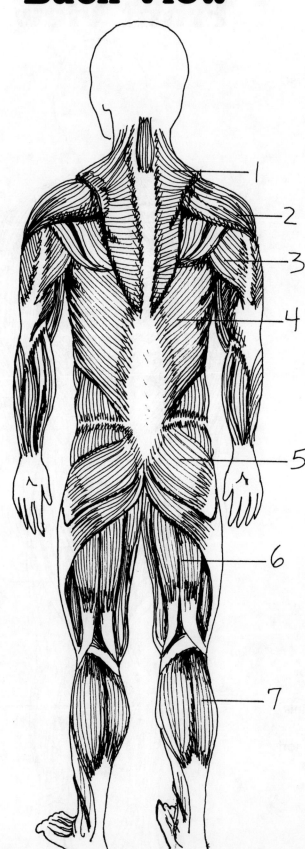

1. Trapezius
2. Deltoid
3. Triceps
4. Latissimus dorsi
5. Gluteus maximus
6. Biceps femoris
7. Gastrocnemius

Muscles Move Bones

To illustrate how muscles move bones, make a muscle and bone model of your arm.

MATERIALS: Stiff cardboard, elastic thread or rubber bands, round top fastener, nail or awl.

PROCEDURE: Use the pattern on the following page to trace the outline of the bones on stiff cardboard. With a nail or awl punch holes at B (biceps muscle), T (triceps), and J (elbow joint). Put the fastener through both holes labeled J connecting the joint. Insert the elastic thread through both holes labeled B and knot. Use another piece of elastic thread to connect both holes labeled T and knot. Observe what happens to the elastic cord as you bend and straighten the elbow joint. You may need to adjust the length of the elastic thread to improve the motion.

QUESTIONS:

1. Which muscle bends the forearm? _____
2. Which muscle extends the forearm? _____
3. Do these two muscles work together? _____
4. Do most muscles work together in pairs? _____

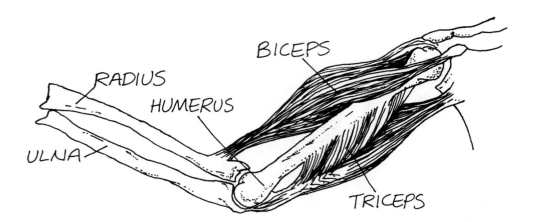

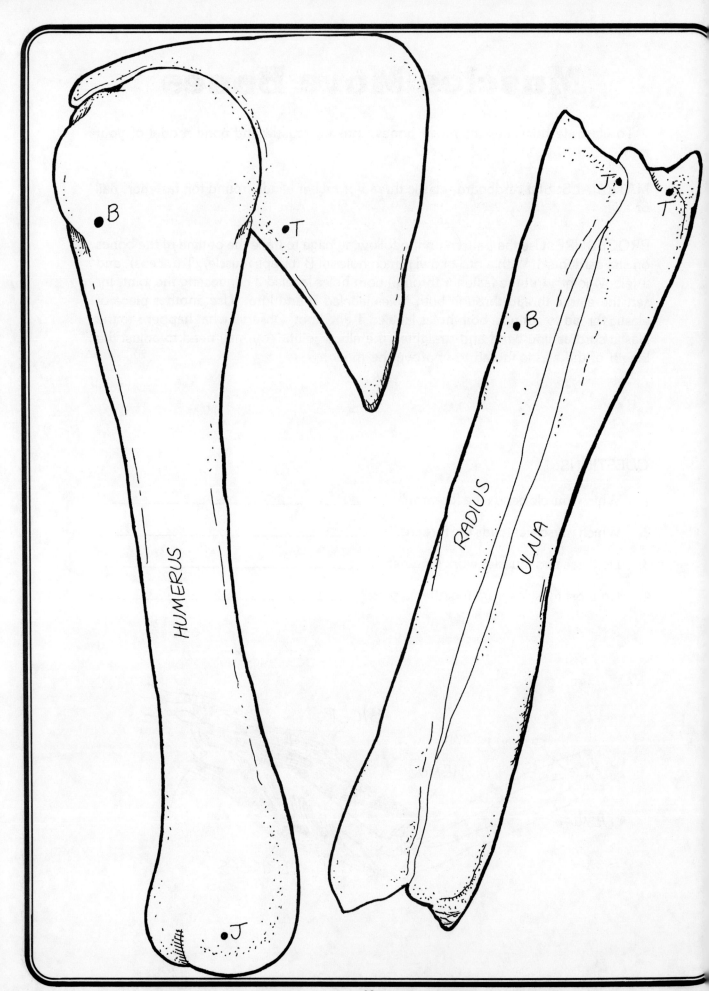

The Circulatory System

The function of the circulatory system is to move or circulate blood. The circulatory system consists of five main parts:
1. Blood, a red liquid.
2. Heart, a muscular pump that moves the blood throughout the body.
3. Arteries, blood vessels which carry blood throughout the body.
4. Veins, blood vessels which return the blood to the heart.
5. Capillaries, tiny blood vessels where the work of the blood is actually done. Through the walls of the capillaries the blood receives the food from the digestive system and the oxygen from the lungs. It is then carried to the individual cells of the body to provide energy in order for them to carry on their functions. The cells give back to the blood in the capillaries carbon dioxide and cellular waste products. The carbon dioxide will be removed in the lungs and the cellular wastes will be taken out when the blood reaches the kidneys.

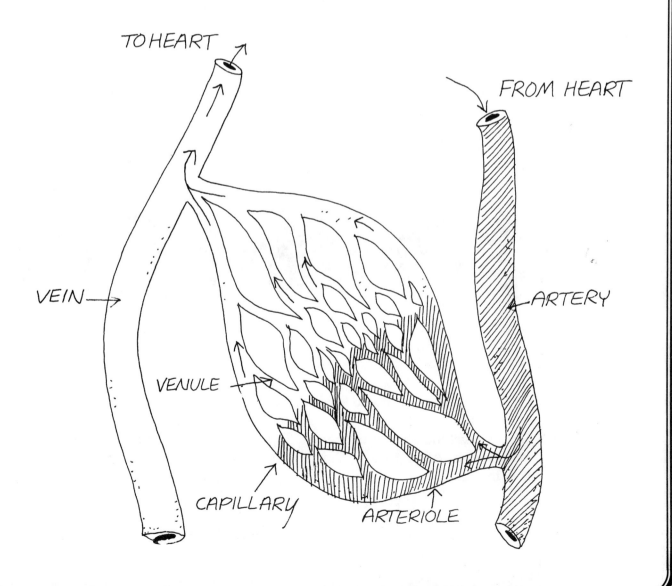

Arteries, Veins, and Capillaries

Arteries are large blood vessels which carry blood away from the heart and lungs. The blood they carry is bright red in color because it has a high content of oxygen in its red blood cells. The walls of arteries are thick and elastic. These walls stretch with each beat of the heart. The arteries branch out into smaller and smaller blood vessels until they become capillaries. Capillaries are very fine blood vessels about one-fiftieth the width of a human hair. Because the walls of the capillaries are so narrow in some places, the blood cells pass through them one at a time.

In the capillaries food and oxygen are given to each individual cell, and cellular wastes and carbon dioxide gas are picked up. As the blood returns to the heart it travels from small veins to larger and larger ones. The walls of the veins are thin, but they have something arteries don't have, valves. Valves are necessary to keep the blood in your legs and other places from flowing backwards when the heart rests between beats.

To show the relationship between arteries, capillaries, and veins, do the following demonstration:

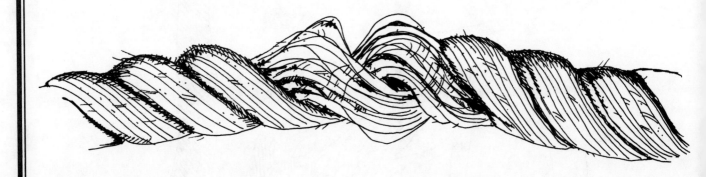

MATERIALS: Rope or heavy cord, red food coloring, blue food coloring, two beakers

PROCEDURE: Unravel the rope in the center. Dip one-half of the rope in a beaker with red food coloring. Allow to dry. Dip the other half in a beaker of blue food coloring. Allow to dry.

Why Is Blood Red?

Blood gets its color from an iron compound in the red blood cells called hemoglobin. Many animals besides humans have red blood, but not all. Lobsters, for example, have pale blue blood because their blood lacks hemoglobin. Our blood is a brighter red when it is in the arteries and full of oxygen. As it gives up its oxygen and travels back to the heart in the veins it turns dark red or purplish. The average 160 pound human body has about five quarts of blood. An interesting way to begin the study of blood is to prepare a demonstration of five quarts of red liquid simulating blood.

MATERIALS: Five one-quart glass or clear plastic containers, red food coloring, water.

PROCEDURE: Make a solution using the red food coloring and water to resemble blood. Fill the containers with the solution. Use this demonstration to begin the study of blood or the circulatory system. To show the amount of blood the heart pumps with each beat, pour 2½ ounces of the red liquid into a measuring cup.

Circulation of Blood Through the Heart

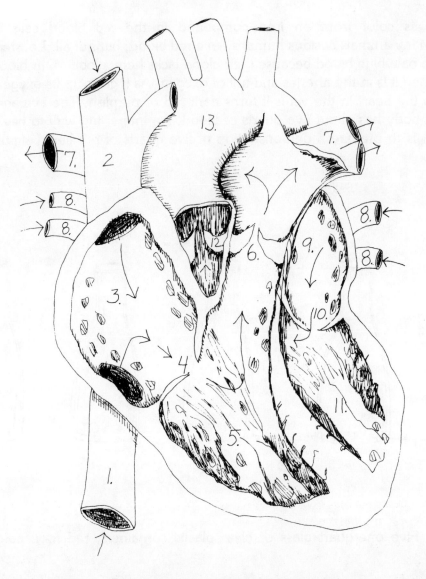

Blood from the lower parts of the body enters the heart through the **inferior vena cava** (1) and from the head and upper parts of the body through the **superior vena cava** (2). These two large veins empty into the **right atrium** (3). The blood then passes through the **tricuspid valve** (4) and into the muscular **right ventricle** (5). Contraction of the walls of the right ventricle forces the blood through the **pulmonary valve** (6) and into the **pulmonary arteries** (7) which lead to the lungs. In the lungs carbon dioxide is exchanged for oxygen and the blood returns to the left side of the heart by way of the **pulmonary veins** (8). It enters the **left atrium** (9) and then passes through the **bicuspid valve** (10) into the largest and strongest chamber of the heart, the **left ventricle** (11). When the left ventricle contracts, the blood is pushed through the **aortic valve** (12) and into the **aorta** (13), the largest blood vessel in the body.

Make a Pulsemeter

Use a pulsemeter to take your pulse.

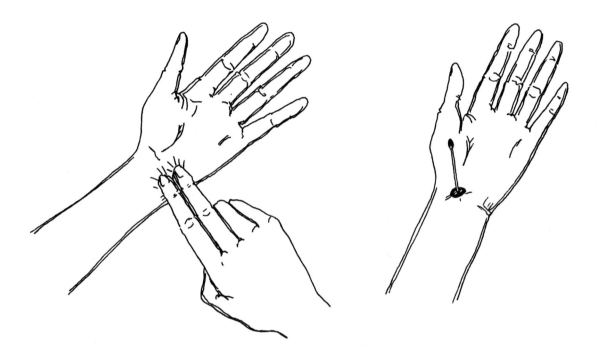

MATERIALS: Thumbtack with a flat round head, wooden matchstick.

PROCEDURE: Carefully insert the thumbtack into the bottom of a wooden matchstick. Set aside. Find the spot on your wrist at the base of your thumb where you can feel your pulse. To take your pulse, use the first two fingers of the opposite hand; do not use your thumb. When you have located a strong pulse, carefully balance the head of the thumbtack on this spot. Your arm should be resting on a desk or table. The top of the match will vibrate with each beat. You may have to move the pulsemeter several times to find the best spot. Keep your arm very still. Take your pulse for one minute with your fingers, and then take it for one minute with your pulsemeter. Were your counts close? Try taking your pulse again after a minute of exercise, such as running in place. Record your results below.

1. Resting pulse taken with fingers _____
2. Resting pulse taken with pulsemeter _____
3. Pulse taken after one minute of exercise _____

VARIATION: Modeling clay and toothpicks also work well.

The Respiratory System

We live on the bottom of an ocean of air which is about twenty percent oxygen. The air we breathe is taken into the body through the nostrils where it is warmed and filtered. It then passes to the back of the throat or pharynx and down the trachea, the tube which leads to the lungs. The walls of the trachea are reinforced with rings of cartilage to keep it from collapsing. Before entering the lungs the trachea forms two branches called bronchi (singular bronchus); each leads to a lung. In the lungs the bronchi branch out like upside down trees; the smallest branches are the bronchioles. At the end of the bronchioles are air sacs or alveoli. It is in the alveoli that the blood receives oxygen and gives up carbon dioxide which is exhaled as a waste product. The oxygen is then carried to all the cells of the body.

Air is drawn into the lungs when the chest is raised and the diaphragm, a sheet of muscle at the base of the lungs, is lowered. This causes a low pressure area which pulls air in. In exhaling air the opposite occurs; the chest is lowered and the diaphragm is raised. Sit quietly and record the number of times you inhale in one minute. Run in place for one minute and count the number of times you inhale. Record below.

Number of Inhalations in One Minute	
Resting	After Exercise

The Respiratory System

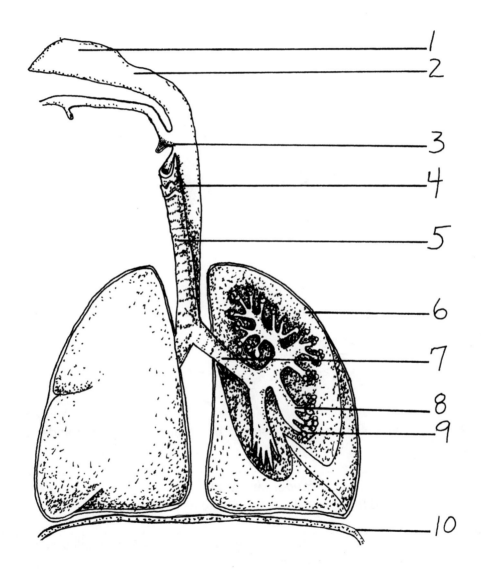

1. Nasal passages
2. Pharynx
3. Epiglottis
4. Larynx
5. Trachea
6. Lung
7. Bronchus
8. Bronchiole
9. Air sac or alveolus
10. Diaphragm

A Model of the Respiratory System

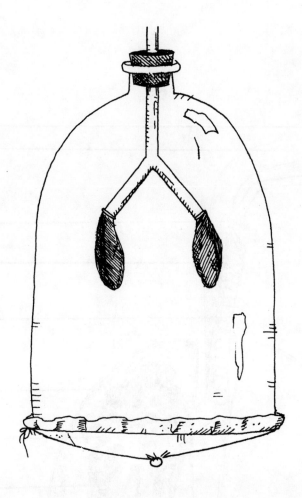

MATERIALS: Bell jar, Y tube, rubber sheet, two balloons, rubber bands, one-hole stopper.

PROCEDURE: Construct the model as shown above. Tell which part of the respiratory system the different parts of the model represent.

Pulling down on the rubber sheet decreases the air pressure in the bell jar and causes air to be drawn into the balloons. Pressing upward on the sheet increases the air pressure in the bell jar and forces the air out of the balloons.

VARIATION: You can build a similar model by using a transparent plastic bottle, one balloon, a plastic drinking straw, clay, a rubber glove and rubber bands. Cut off the bottom of the plastic bottle and punch a hole in the cap for the straw. Attach the balloon to the straw with a rubber band and insert through the hole in the cap. Seal any spaces around the straw with clay. Stretch the flat part of a rubber glove over the base of the bottle with rubber bands. Use as you would the model pictured above.

How Much Air Can Your Lungs Hold?

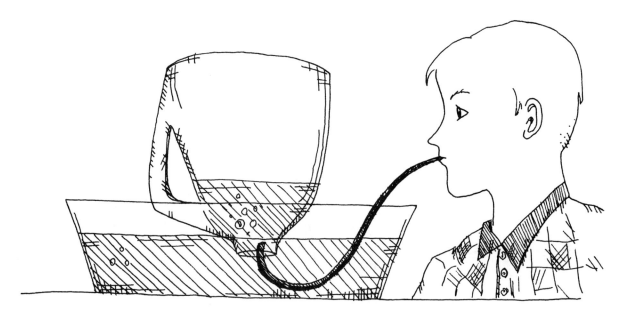

MATERIALS: One-gallon plastic jug, rubber or plastic tubing, very large pan or aquarium, water.

PROCEDURE: Fill the gallon jug with water and cap. Pour several inches of water into the pan. Turn the jug upside down and place in pan below the water level and remove the cap. Insert one end of the tubing into the mouth of the jug. Take one deep breath and exhale into the other end of the tubing until you have emptied all the air out of your lungs. Cap the jar and remove from the pan. With a magic marker mark your lung capacity on the jug. You may wish to measure the remaining water with a measuring cup or beaker to find exactly how much water was replaced by air from your lungs.

Note: This experiment may be coordinated with the one on the following page. To do so, save materials and results.

How Much Oxygen Is Left After Breathing?

By doing this experiment you will get an idea of how much oxygen is left in the air exhaled from your lungs.

MATERIALS: Milk jug with cap, candle, matches, coat hanger or stiff wire, watch or clock with second hand, materials from preceding experiment.

PROCEDURE: Fill one jug with water, then pour the water out. The jug will automatically fill with fresh air. Use the wire to make a handle for the candle. Light the candle and quickly insert it into the jug and cover. As soon as the jug is covered, keep track of the time it takes for the candle to go out. Record the time. The candle needs oxygen to burn; as soon as the oxygen is used up it will go out. Repeat the first part of the experiment on the preceding page, but this time keep breathing until all the air has been emptied from the jug. (You are not measuring lung capacity now, so you may take as many breaths as needed to empty the jug.) When all the air has been emptied from the jug, cap it and remove from the pan. Light the candle and remove cap from the second jug. Insert the candle quickly and cap. As soon as the jug has been capped, keep track of the time it takes for the candle to go out. Record the time.

Amount of time candle burns in fresh air _____

Amount of time candle burns in exhaled air _____

The Digestive System

    Digestion is the process of breaking food into small particles which can be used by the cells of the body for energy, growth and reproduction. The four basic types of food acted upon in digestion are fats, proteins, sugars and starches. Digestion begins in the mouth when saliva secreted by three pairs of salivary glands start the breakdown of starches into simple sugars. The saliva-moistened food then travels down the esophagus in lumps called boli to the stomach. As the muscular walls of the stomach churn and mix the food, the gastric glands secrete enzymes that begin the digestion of proteins. The partially digested food leaves the stomach through the pyloric valve and enters the twenty-two-foot-long small intestine. Ducts from the liver, gall bladder and pancreas lead to the upper part of the small intestine; the liver and the gall bladder add bile to breakdown fats while the pancreatic juices act on proteins. The food is now ready to be absorbed in the many tiny, fingerlike projections in the small intestine called villi. Here the blood picks up digested food and carries it to all parts of the body. Undigested food moves on to the six-foot-long large intestine where some water is removed and waste is stored before leaving the body through the rectum. The appendix is not needed for digestion and performs no necessary function in modern man.

The Digestive System

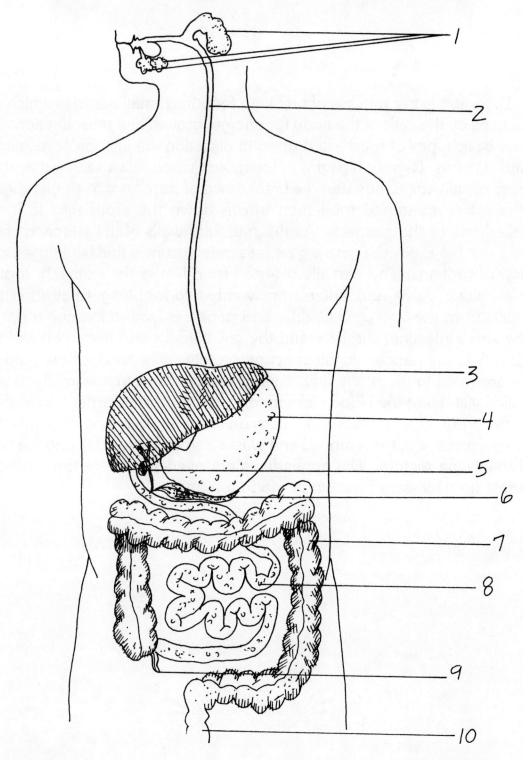

1. Salivary glands
2. Esophagus
3. Liver
4. Stomach
5. Gall bladder
6. Pancreas
7. Large intestine
8. Small intestine
9. Appendix
10. Rectum

Breaking Down Food for Digestion

To show how the breaking down of food particles aids in the digestion of foods, do the following demonstration:

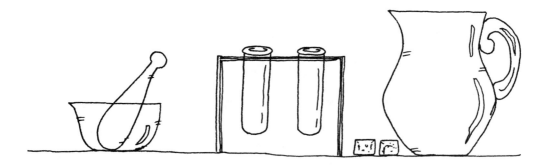

MATERIALS: Two test tubes, test tube rack, two sugar cubes, mortar and pestle or spoon or cup, water.

PROCEDURE: Crush one cube of sugar with the mortar and pestle or by using the spoon and sides of the cup. Put the powdered sugar cube into a clean dry test tube and the sugar cube in the remaining test tube. Fill both test tubes about two-thirds full of water. Observe what happens to the sugar in the test tubes.

QUESTIONS:

1. In which test tube did the sugar dissolve more quickly? Why?

2. How is the breaking of the sugar cube similar to the chewing of food?

3. Where does digestion begin?

Teeth

The main function of teeth is to break food down into small pieces so that it can be more easily mixed with the digestive juices. Man has two sets of teeth. The first set, the deciduous or baby teeth, appears several months after birth and are later replaced by permanent teeth. The front teeth are the incisors; they are shaped for cutting and biting. The cuspids and bicuspids have points for tearing food and the molars have flat surfaces for grinding.

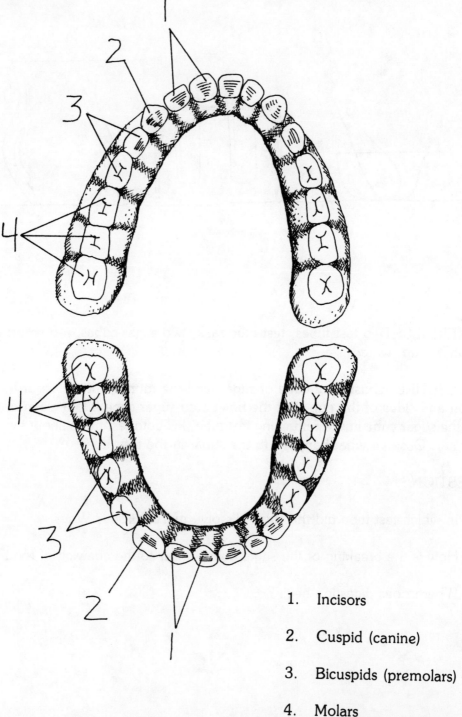

1. Incisors
2. Cuspid (canine)
3. Bicuspids (premolars)
4. Molars

The Parts of a Tooth

The hardest part of a tooth is its covering, the enamel. Beneath the enamel is a substance similar to hard bone called dentine. Most of the tooth is made of dentine. The cementum fastens the root of the tooth to the jawbone. The central part of the tooth is called the pulp cavity; it contains the blood vessels and nerves which enter the tooth through small openings at the tips of the roots and travel up the root canal.

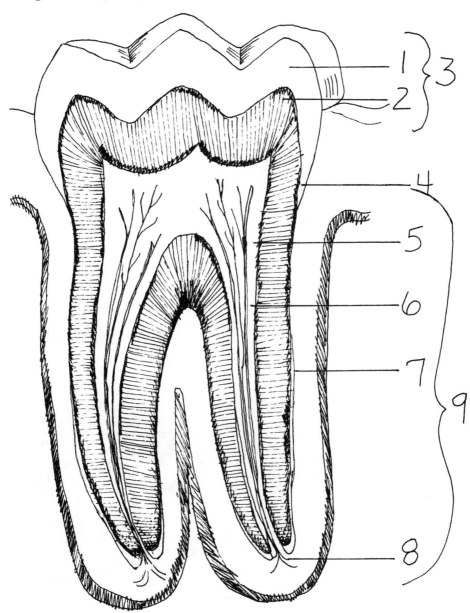

1. Enamel
2. Dentine
3. Crown
4. Neck
5. Pulp
6. Root canal
7. Cementum
8. Nerves and blood vessels
9. Root

The Excretory System

The internal organs for the removal of liquid waste or urea from the blood are kidneys. They are located in the small of the back just below the diaphragm and are protected by the lower ribs. The kidneys act as filters which take chemicals and other poisons or toxins out of approximately fifty quarts of blood a day. These waste products are removed by collecting tubules. The liquid waste, now called urine, leaves each kidney through a ureter. The ureters lead to the bladder which is a storage organ for urine. The bladder is emptied by a single tube leading to the outside of the body called the urethra.

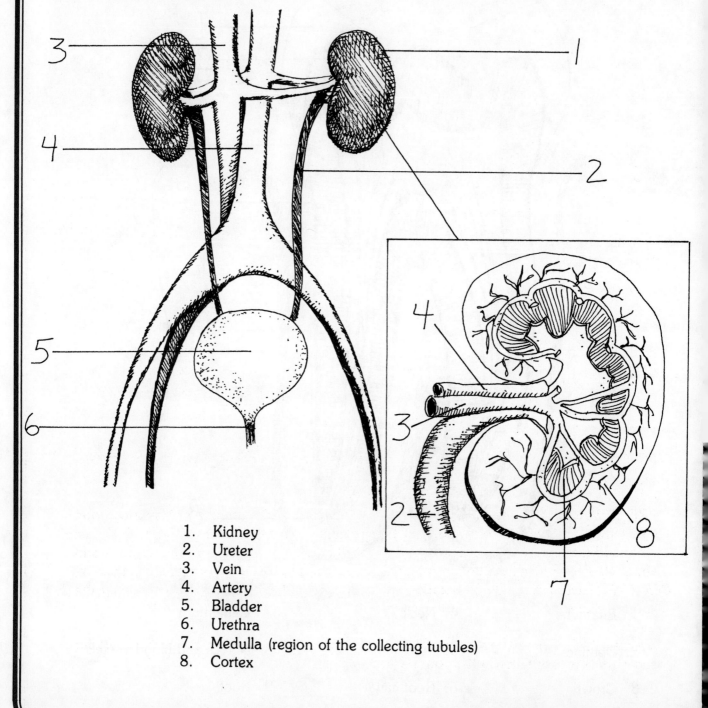

1. Kidney
2. Ureter
3. Vein
4. Artery
5. Bladder
6. Urethra
7. Medulla (region of the collecting tubules)
8. Cortex

A Test for Diabetes

Special cells in the pancreas regulate the amount of sugar (glucose) in the blood. When these cells are not functioning properly there is an increase in the amount of sugar secreted in the urine. This increase in sugar may be an indication of a disease called diabetes.

One way to test for sugar is to use Fehling's solution. To show how sugar can be detected in urine, do the following experiment:

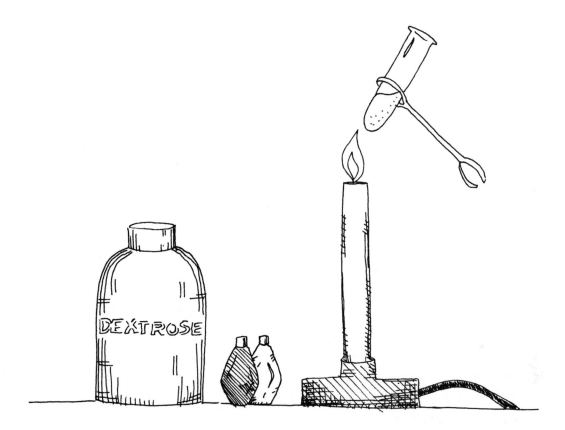

MATERIALS: Two test tubes, water, yellow food coloring, Bunsen burner, glucose or dextrose powder, Fehling's A and B.

PROCEDURE: Fill the test tubes one-third full of water. Add a drop or two of yellow food coloring to simulate the color of urine. Label the test tubes A and B. Into one of the test tubes add a small amount, about ¼ teaspoon, of glucose powder. Into each test tube add about 10 cc of Fehling's A and B solutions. Heat carefully with the test tube pointed away from the face. The Fehling's solution will turn brick red in the test tube containing the "urine" with sugar.

VARIATION: Clinitest tablets available at most drugstores can also be used to test for sugar. Follow instructions given on the package carefully.

The Nervous System

The human nervous system is made up of the brain, spinal cord and nerves of the body. The brain and spinal cord are called the central nervous system. The other nerves are called the peripheral nervous system. One of the jobs of the peripheral nervous system is to supply information to the brain.

The brain is protected by the skull, three membranes and a special fluid. In man the largest part of the brain is the cerebrum. The cerebrum, in addition to being the thinking portion of the brain, receives and interprets information from the five senses, controls emotions, speech and memory. The cerebellum is in charge of muscular coordination and balance. The third main portion of the brain, the medulla, controls the involuntary workings of internal organs such as the beating of the heart, breathing and digestion.

The spinal cord connects the brain to the peripheral nerves. It is protected by the bones of the back called vertebrae. Holes in the vertebrae allow the nerve messages to be carried to the spinal cord where they are usually transferred to the brain for decisions. Sometimes decisions are made without the help of the brain; this is called a reflex act because the message only goes as far as the spinal cord. An example of a reflex act is the sudden lifting of your hand when you touch a hot stove. Before you have time to think, the muscles of your arm quickly react to remove your hand from the stove. Reflex acts allow you to react first and think later thereby saving valuable time in dangerous situations.

The Nervous System

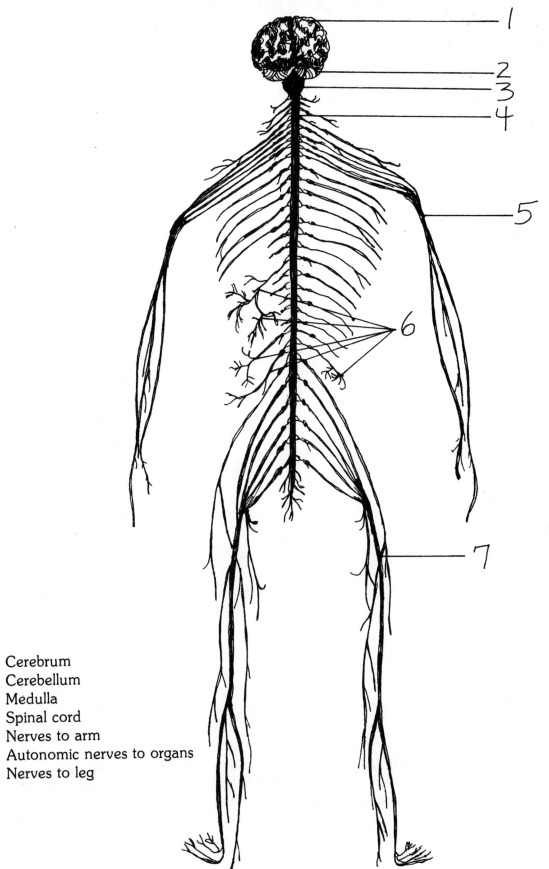

1. Cerebrum
2. Cerebellum
3. Medulla
4. Spinal cord
5. Nerves to arm
6. Autonomic nerves to organs
7. Nerves to leg

The Brain

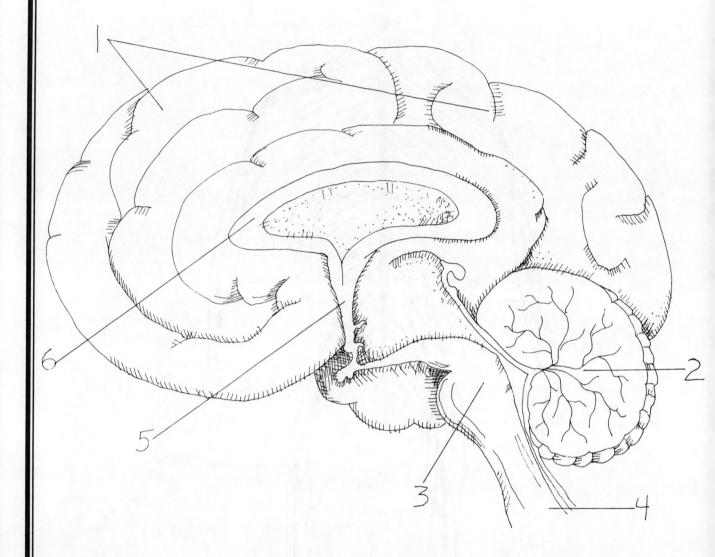

1. Cerebrum - this is the intelligence center of the brain. It controls memory, thinking, learning, etc. It also receives and interprets messages from the five senses.
2. Cerebellum - controls muscular coordination and balance.
3. Medulla - controls involuntary functions such as breathing, heartbeat, digestion.
4. Spinal cord - this is the large nerve leading from the brain and extending down the back.
5. Hypothalamus - this is the control center for appetite, water balance, sleepiness and temperature.
6. Thalamus - this is the center for anger, pleasure and basic drives.

Our Sense of Sight

Our best and most valued sense is that of sight, but in order to see there must be light. Animals that live in deep caves or in the great depths of the ocean where there is no light are often blind or have no eyes at all.

We see when light waves strike an object which we are viewing and are bounced off the object and reflected into our eyes. When these reflected light waves reach our eyes, they must first pass through a delicate layer of cells called the cornea (kor-ne-ah). From the cornea the light waves pass through an opening into the eye called the pupil. The pupil is the black dot you see in the middle of your eye; it is surrounded by the colored part of the eye, the iris. In back of the pupil is the lens. The lens brings the viewed object into focus and also turns it upside down. The inside of the eyeball is called the retina (ret-nah). The retina is made of cells called rods and cones because of their shapes. These cells are sensitive to the light waves, and it is here that a picture of what we see is actually formed upside down. This picture is carried to the brain by the optic nerve where it is turned rightside up and understood with help from other parts of the brain such as the memory section.

ACTIVITY: List the five senses on the chalkboard. Have the class vote on which sense would be the most difficult to give up?

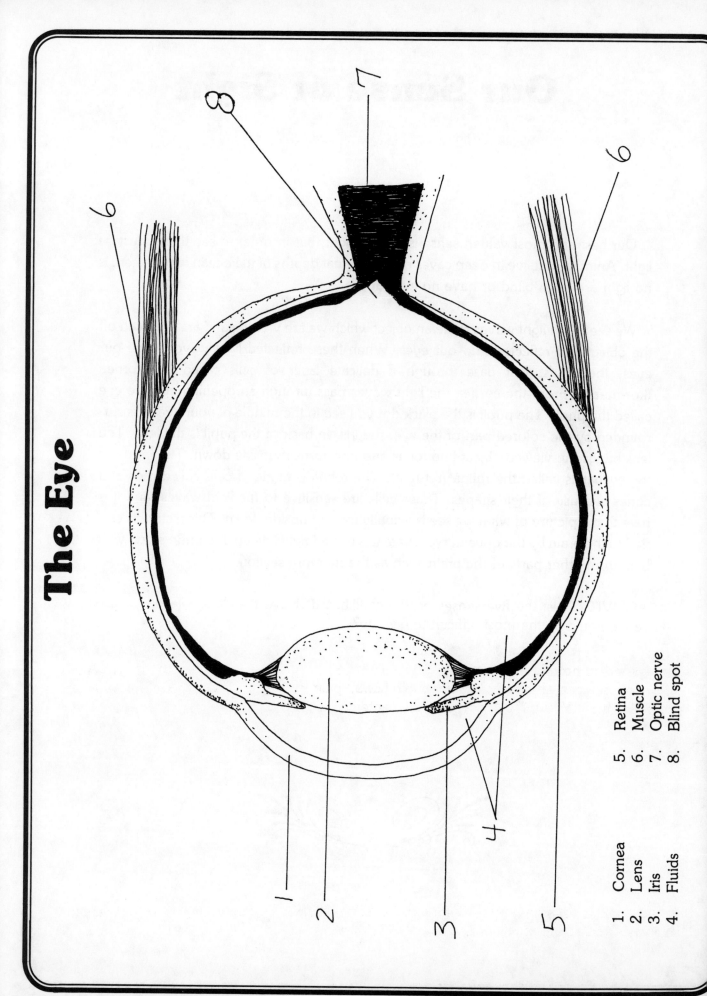

Eye Tricks

Make your eyes play tricks on you; try the following optical illusions:

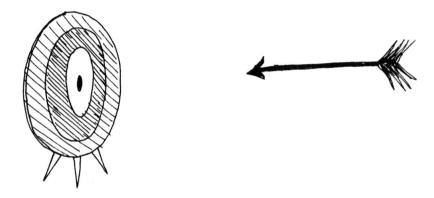

To have the arrow hit the target put the picture at eye level and slowly bring it in toward your eyes. Design a similar optical illusion of your own. Some suggestions: A dog catching a bone, a glove catching a baseball.

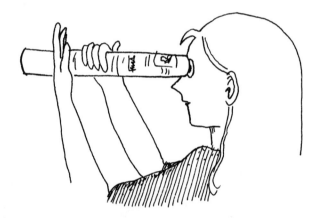

Roll a sheet of notebook paper lengthwise. Hold it up to your right eye. Keep both eyes open and look straight ahead. Place your left hand, palm facing you, next to the middle of the paper tube. You will see a hole in the center of your hand.

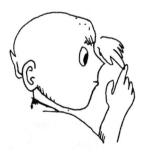

You can make a sausage appear before your eyes if you bring your index fingers together about six inches away from your nose.

Test Your Stereoscopic Vision

Our two eyes working together give us stereoscopic vision. Stereoscopic vision is very important in the judging of distances, as in driving a car. Loss of sight in one eye would require much practice in learning to judge distances accurately, something most of us take for granted. To demonstrate how both eyes work together to enable us to judge distances, do the following experiment:

MATERIALS: A small test tube and a colored plastic toothpick for each pair of students.

PROCEDURE: One student will hold the test tube at arm's level about 24 inches away from his partner. The other partner will cover one eye and attempt to drop the toothpick into the test tube. (Touching the tube before dropping the toothpick is not allowed.) Try five times with each eye; partners then reverse roles.

Our Sense of Hearing

Human ears, although not as sensitive as those of many other animals, are capable of receiving and transmitting high frequency vibrations (sound waves).

The human ear is divided into three parts: the outer ear, the middle ear and the inner ear. The outer ear consists of the auricle and the auditory canal. The middle ear consists of the eardrum, and three small bones. The inner ear contains the cochlea, the three semicircular canals and the auditory nerve.

We hear when vibrating or moving objects send out sound waves which travel through the air to our ears. The shell-like shape of the outer ear (auricle) catches the sound waves and reflects them into the auditory canal which leads to the eardrum. The sound waves cause the delicate eardrum to vibrate, and it, in turn, causes the vibration of three tiny bones: the hammer, the anvil and the stirrup. The last bone, the stirrup, transfers the vibrations to the cochlea, a snail-shaped organ filled with liquid. The motion of the fluid starts impulses in the sensory nerves leading to the auditory nerve. The auditory nerve relays these impulses to the brain when they are interpreted.

In the inner ear the three small semicircular canals serve as an organ for balance. Each canal is located on a different plane and is filled with liquid. When the head is moved, the liquid in the canals also moves causing small hair-like projections in the canals to move. These projections are connected to nerves which send the messages of movement to the brain. The brain on receiving these messages directs the muscles of the body to maintain balance.

The Ear

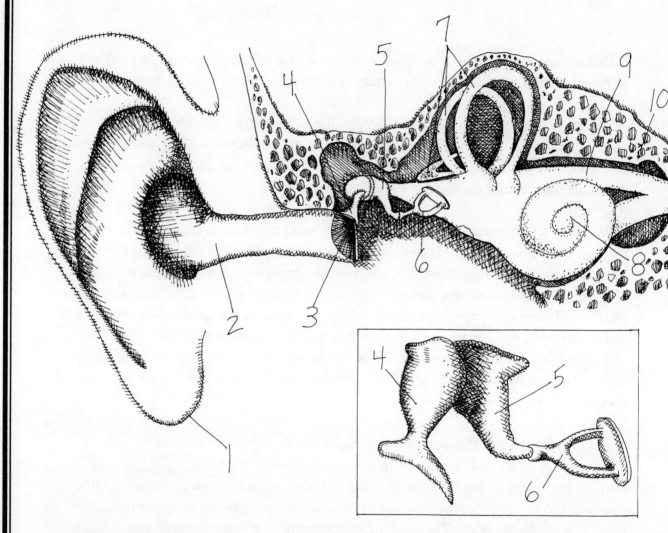

1. Auricle or pinna
2. Ear canal (auditory canal)
3. Eardrum
4. Hammer (malleus)
5. Anvil (incus)
6. Stirrup (stapes)
7. Semicircular canals
8. Cochlea
9. Auditory nerve
10. Bone

Using One Ear to Detect Sounds

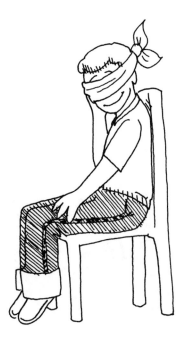

MATERIALS: A blindfold, noisemakers, such as wooden blocks, empty tin can, wooden spoons, coins, etc.

PROCEDURE: Select a student and seat him in the center of the room with eyes closed or blindfolded. All other students must remain very quiet. Select three or four other students as noisemakers. Provide the noisemakers with objects that will make interesting noises such as wooden blocks, wooden spoons, two coins, empty cans, etc. Have the noisemakers stand at various spots 12 to 15 feet away from the subject. The subject should not know beforehand the objects used to make the sounds. The subject should now cover one ear. The noisemakers will make tapping sounds, one at a time. After each noise the subject will point to the direction from which he thinks the sound originated. Have the subject test his opposite ear. Move the noisemakers quietly to other positions.

VARIATIONS:

1. Try the experiment with both ears uncovered.

2. Try the experiment with both ears covered.

Our Sense of Touch

In our skin are special sensory cells that pick up messages from the outside world called receptors. Some receptors tell us when we touch something or feel pressure; others make us aware of heat, cold and pain. The skin also protects the internal organs of the body from injury and germs, helps to regulate the body's temperature and excretes sweat, a liquid waste.

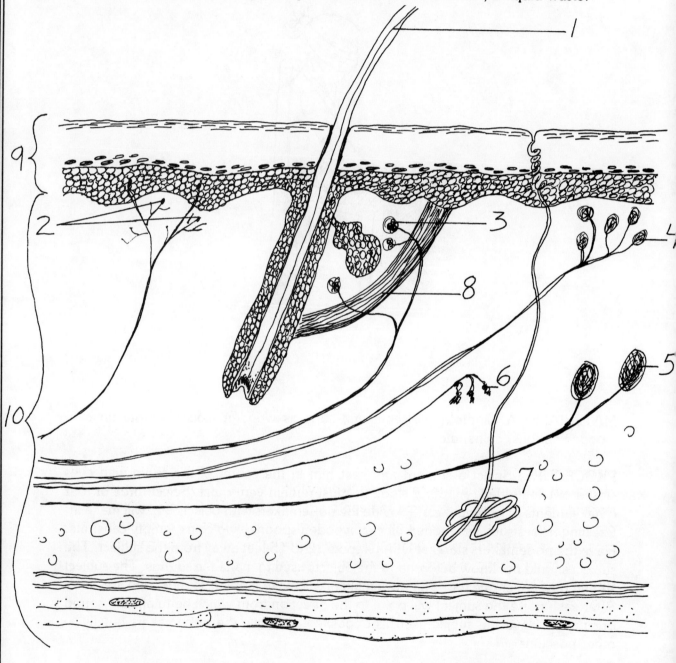

1. Hair
2. Pain receptors
3. Cold receptor
4. Touch receptor
5. Pressure receptor
6. Heat receptor
7. Sweat gland
8. Muscle
9. Epidermis
10. Dermis

How Well Do You Feel?

To test your sense of touch and memory do the following activity:

MATERIALS: Rocks.

PROCEDURE: Ask each student to bring to class a rock that can be concealed in the palm of his hand. Have the students study their rocks, and then with permanent magic markers mark their rocks with identifying symbols or their initials. Take up the rocks. Do not tell students what they are going to do with the rocks. The next day arrange the students in a circle and ask them to keep their eyes closed for the experiment. Randomly distribute the rocks. At a given signal have students pass the rocks from left to right. When a student thinks he has received his rock, he drops out of the experiment with his rock. Continue until all the rocks are claimed or the experiment comes to a standstill. Have students open their eyes. How many claimed the correct rock by using only the sense of touch?

Hot or Cold?

MATERIALS: Three basins: one with hot water, one with cold water and one with warm water.

PROCEDURE: This little experiment shows that it is not always a simple matter to tell the difference between water temperatures while blindfolded. Sometimes our sense of hot and cold can play tricks on us. First fill three basins with water, one with hot (not too hot, of course!), one with cold, and the third with lukewarm. Have the basins arranged so that the student can place his right hand in the hot water, his left hand in the cold. Allow hands to soak for a few minutes. Then ask the student to plunge both hands into the middle bowl of lukewarm water. He will discover that the right hand feels cold, and the left hand feels warm!

Our Sense of Taste

    Have you ever noticed the large and small bumps on your tongue? In these bumps are your taste buds. We taste the flavors of so many varieties of foods that you probably think there are thousands of different kinds of taste buds, but there are really only four: sweet, sour, salty and bitter. The taste buds are not scattered and mixed over the surface of the tongue, but are found only in certain parts. Taste buds which detect sweet foods are found along the front and sides of the tongue; salty foods are also tasted along the front and sides while sour foods are tasted only along the sides. Bitter substances can only be tasted at the back as you may have noticed if you have ever swallowed bitter medicine.

In the drawings below, shade in the particular area where each type of taste bud is found.

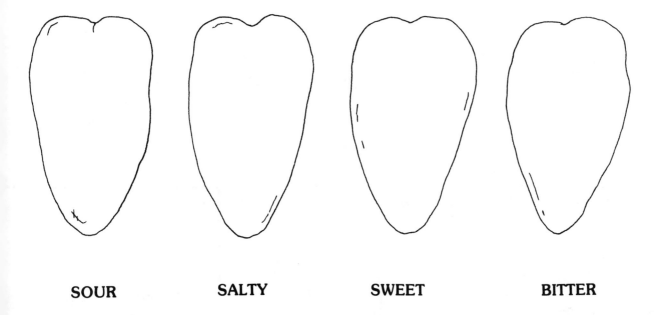

SOUR   **SALTY**   **SWEET**   **BITTER**

Rate the Taste

Do most people like to eat the same things? To find out how students rate certain foods in comparison to their classmates, have students rate the following foods on a scale of 1 to 10. Tally the scores. This would be a good time to have students practice using calculators if they are available. Put the results on the chalkboard.

INSTRUCTIONS: Circle the number to rate each of the following foods; number 10 is the highest rating.

Our Sense of Smell

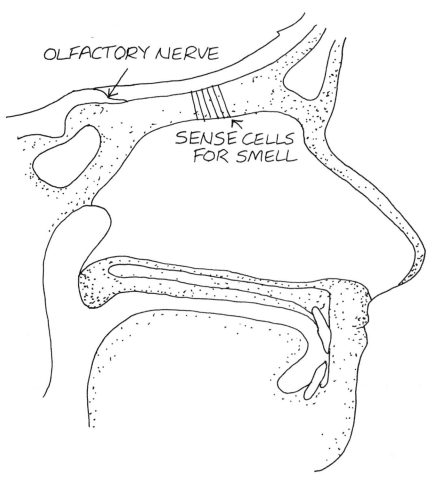

The nose protects a small patch of cells located on the roof of the nasal cavity. These cells can detect, on the average, about two thousand different odors. The message from these odors is relayed to the brain by means of a special nerve called the olfactory nerve. The brain interprets these odors and tells us how to react to them. Some odors are pleasant; others are not. Below are some common odors; rate them as being pleasant, unpleasant, or neither. Compare ratings. Do boys and girls find the same odors pleasant?

1. Coffee
2. Cigar
3. Apple pie
4. Newly mown hay
5. Onion
6. Christmas tree
7. Spaghetti
8. Gasoline
9. Pickle
10. Vanilla

Is It Taste or Smell?

Much of what we think we taste is actually the smell of food. Your class will surely enjoy this experiment which shows this fact to be correct.

MATERIALS: Slices of orange, slices of potato, slices of onion, slices of apple, and slices of various other foods or fruits in season such as melon, strawberries, pears, etc., a blindfold, a volunteer taster who likes all kinds of foods.

PROCEDURE: Have the volunteer seated comfortably and blindfolded. The volunteer may also close his eyes, and he should not have seen the foods beforehand. The volunteer must also hold his nose firmly while he is tasting. Ask the volunteer to stick out his tongue. Rub a slice of the mystery food on his tongue. Keep a record on the chalkboard of the foods used and the volunteer's responses. After all the foods have been sampled, tally the number of correct and incorrect responses. The experiment may be repeated with the volunteer blindfolded but not holding his nose.

Warning Smells

Has your sense of smell ever warned you of possible danger? Below are some examples of dangers you may not always see or hear. Tell how your sense of smell can make you aware of what is happening in each case. Try using some of these situations in a cartoon strip of your own to show times when you rely on your sense of smell to come to your aid.

Anatomy of the Mouth, Nose and Throat

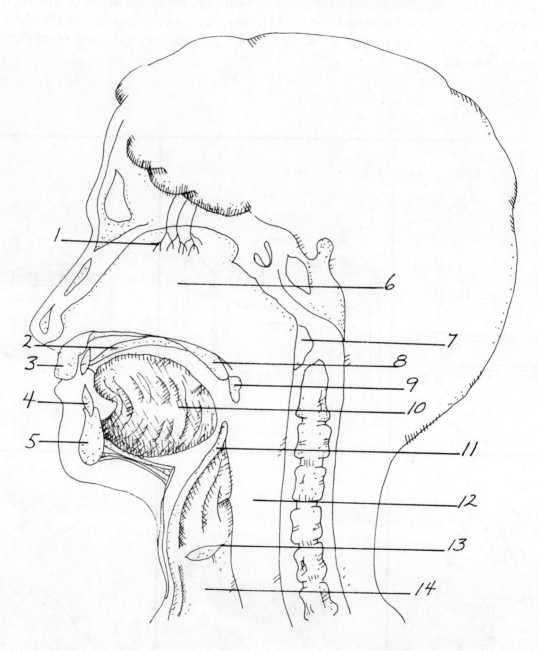

1. Nerves of smell (olfactory)
2. Hard palate
3. Upper lip
4. Tooth
5. Lower jaw (mandible)
6. Nasal passage
7. Adenoids
8. Soft palate
9. Uvula
10. Tongue
11. Epiglottis
12. Esophagus
13. Vocal cord
14. Windpipe (trachea)

The Female Reproductive System

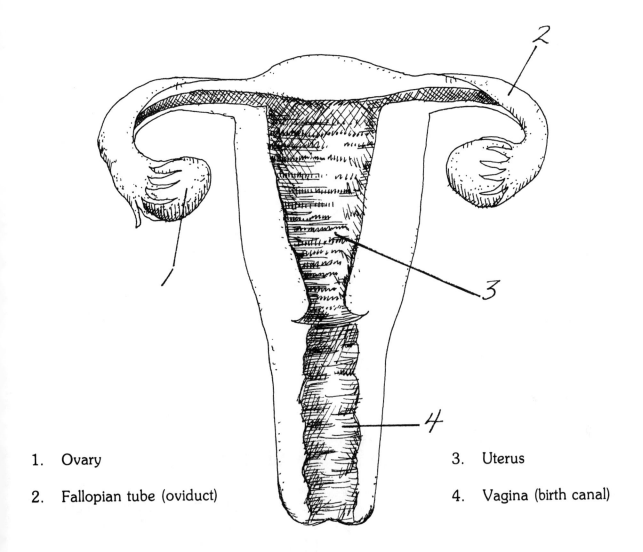

1. Ovary
2. Fallopian tube (oviduct)
3. Uterus
4. Vagina (birth canal)

In humans the female reproductive cell is called the egg or ovum. It is produced in an organ called the **ovary.** The human female has two ovaries. They are a little less than an inch and a half long and a half inch wide. They are located one on each side of the lower part of the back. An egg is produced from an alternating ovary approximately once every twenty-eight days. When the mature ovum is released from the ovary it is picked up by the **Fallopian tube** or oviduct. It travels along the Fallopian tube until it reaches a pear-shaped, muscular organ called the **uterus.** If the egg has not been fertilized, it disintegrates and leaves the body along with some blood and uterine tissue through the **vagina.** This discharge of blood and tissue is called menstruation.

The Male Reproductive System

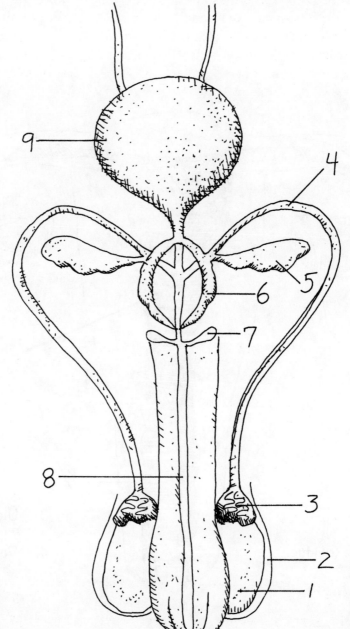

1. Testis
2. Scrotum
3. Epididymis
4. Vas deferens
5. Seminal vesicle
6. Prostate gland
7. Cowper's gland
8. Urethra
9. Bladder

　　The male reproductive organs are called **testes.** They are located on the outside of the body in a pouch called the **scrotum.** The production of sperm is the most important function of the testes. From the tubules of the testes the sperm travel to the tubules of the **epididymis** (ep-uh-DID-uh-miss) where they are sometimes stored. The **vas deferens** carries the sperm from the epididymis to the **urethra** which leads to the outside of the body. Secretions are added to the sperm in several places mainly in the **Cowper's glands, seminal vesicles,** and the **prostate glands.** The sperm together with the secretions of the male reproductive organs are called semen.

The Endocrine System

The endocrine glands secrete complex chemical substances called hormones directly into the bloodstream. These hormones maintain a delicate chemical balance in the body. When an endocrine gland does not function properly, it is because it makes too much or too little of its hormone. Oversecretion of a hormone is called hypersecretion; too little of a secretion is called hyposecretion. Hormones are so powerful that even a small change in secretion can have very serious results.

1. Pituitary
2. Thyroid
3. Parathyroid
4. Thymus

5. Adrenal
6. Pancreas
7. Ovary
8. Testis

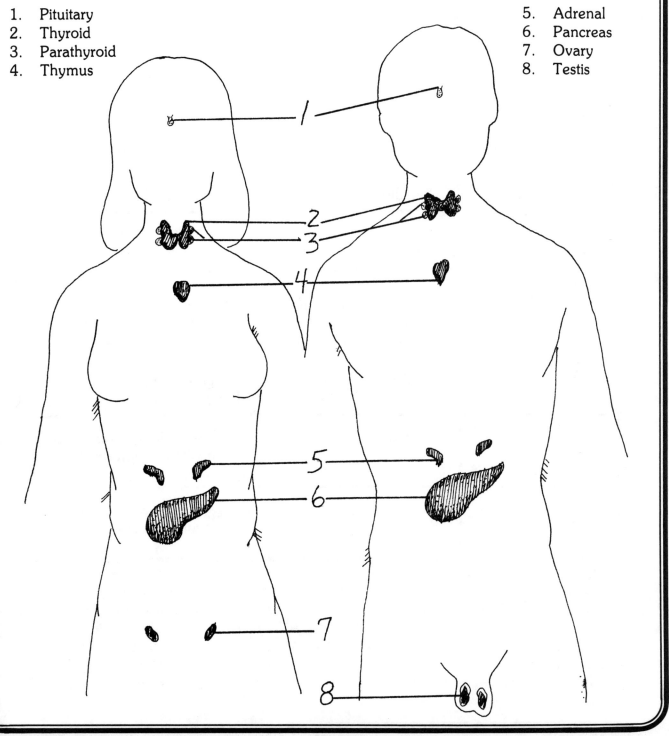

Endocrine Glands

Write the name of the gland in the first column that matches the major secretion(s) and functions of that gland.

Endocrine Gland	Major Secretion(s)	Functions
	ACTH and many complex hormones	This "master" gland produces hormones which regulate growth and metabolism by controlling the other glands in the endocrine system.
	Thyroxin	Regulates the rate at which the body burns food (metabolism).
	Parathormone	Controls calcium and phosphorus levels.
	Thymosin	Controls immunities in children.
	Adrenalin, cortisone, and others	Stimulates the nervous system during emergencies.
	Insulin	Regulates the rate at which cells burn up sugar in blood.
	Estrogen	Regulates female sex development and reproduction.
	Testosterone	Regulates male sex development and reproduction.

Answer Key

Page 1: Systems of the Body

System	Major Organs	Major Functions
Skeletal	bones of the body	support protection of organs
Muscular	muscles of the body	movement
Circulatory	heart, arteries, veins, capillaries, blood	carry food and oxygen to cells, return waste products for excretion by kidneys
Respiratory	nasal passages, trachea, bronchi, lungs	exchange of gases
Digestive	esophagus, stomach, small intestine, large intestine, etc.	break foods down so that they may be absorbed by the blood
Excretory	kidneys, ureters, bladder, urethra	get rid of liquid wastes
Nervous	brain, spinal cord, nerves	body controls
Reproductive	ovaries in female, testes in male	production of eggs and sperm
Endocrine	pituitary, thyroid, parathyroid, thymus, adrenals	maintain balance for growth, metabolism and reproduction

Page 5: What Is Bone?

1. The bone in water remained the same.
2. The bone in acid became soft.
3. Salts
4. It burned.
5. The protein
6. Living cells get their nourishment from the kinds of food we eat.

Answer Key

Page 9: Determine the Fracture

1. Ricky's
2. Fuller's
3. Todd's
4. Karen's
5. Mary's
6. Fred's

Page 11: Joints

1. Ball and socket (hip)
2. Hinge (elbow)
3. Immoveable (skull)
4. Ball and socket (shoulder)
5. Slightly moveable (vertebrae)
6. Hinge (knee)
7. Gliding (wrist)
8. Pivot (atlas and axis bones in neck)

Page 12: The Muscular System

1. b
2. e
3. d
4. a
5. c
6. f

Page 15: Muscles Move Bones

1. Biceps contracting
2. Triceps contracting
3. Yes
4. Yes

Page 29: Breaking Down Food for Digestion

1. The one in which the sugar was crushed. Smaller pieces are easier to dissolve.
2. Food is crushed and broken down by the teeth.
3. In the mouth when the food is broken into smaller pieces and is mixed with saliva.

Page 51: Warning Smells

a. Food spoiling
b. Bits of food on burner warning that stove is on
c. Food burning
d. Fire
e. Skunk
f. Odors escaping from spilled acid

Page 56: Endocrine Glands

Glands in order, top to bottom are:

Pituitary
Thyroid
Parathyroid
Thymus
Adrenals
Pancreas
Ovaries
Testes